Samuel Haughton

Address on the relation of food to work and its bearing on medical practice

1868

Samuel Haughton

Address on the relation of food to work and its bearing on medical practice
1868

ISBN/EAN: 9783337201449

Printed in Europe, USA, Canada, Australia, Japan

Cover: Foto ©berggeist007 / pixelio.de

More available books at **www.hansebooks.com**

ADDRESS

ON THE

RELATION OF FOOD TO WORK,

AND

ITS BEARING ON MEDICAL PRACTICE:

DELIVERED

BEFORE THE BRITISH MEDICAL ASSOCIATION, IN THE DIVINITY SCHOOL AT OXFORD, ON THE FIFTH OF AUGUST, 1868,

BY

THE REV. SAMUEL HAUGHTON, F.R.S.,

M.D. DUBL., D.C.L. OXON.,

FELLOW OF TRINITY COLLEGE, DUBLIN.

"Ridiculum acri
Fortius et melius magnas plerumque secat res."

DUBLIN

PRINTED AT THE UNIVERSITY PRESS,

BY M. H. GILL.

1868.

ADDRESS

ON THE

RELATION OF FOOD TO WORK,

AND

ITS BEARING ON MEDICAL PRACTICE,

&c. &c.

Mr. President and Gentlemen,

Man, like other animals, is born, grows, comes to maturity, reproduces his like, and dies; passing in his lifetime through a cycle of changes that may be compared to a secular variation, by a metaphor borrowed from the science of Astronomy, while, in his daily life, he passes through a smaller cycle of changes that may be called periodic.

From the time of the publication of Bichat's celebrated Essay on Life and Death, it has been admitted that man and other animals possess a double life, animal and organic, presided over respectively by two distinct, though correlated centres of nervous force; of these, one thinks, moves, and feels; the other merely cooks; receiving the food supplied, changing and elaborating it into elements suitable for the use of the animal life. In the lower forms of animals, the organic life becomes almost coextensive with the whole being of the creature, which simply digests, assimilates, and excretes; but barely feels or moves. In the higher forms of animals, and more especially in man, the animal life dominates over the organic life, which becomes its slave, and exhibits the remarkable phenomena of mechanical force, of geometrical instinct, of animal cunning, and finally, in man himself, produces intellectual work, rising to its highest form in the religious feeling that recognizes its great Creator, and bows in humility before Him. It is a simple matter of fact, and of every day observation, that all these forms of animal work are the result of the reception and assimilation of a few cubic feet of oxygen, a few ounces of water, of starch, of fat, and of flesh.

The general question of the relation of Food to Work would involve a consideration of the possibility of throwing a bridge across the gulf that separates the organic from the animal life, so as to connect the products of nutrition (taken in its widest sense) with the work of every kind accomplished by the animal life, whether mechanical or intellectual. We resemble the spiders of the heather on a summer morning, that float their gossamer threads into the air from the summit of a branch, in the hope that some stray breath of wind may fasten them to a neighbouring tuft, and enable the hungry speculator to extend the range of his rambles and his chance of food. Already a few feeble threads connect the chemistry of our food with the mechanical work done by our muscles; when these shall have been securely fastened, from the higher vantage ground thus acquired, our little bridge of knowledge may possibly be extended to embrace the phenomena of the geometrical instinct of the bee, or the cunning of the beaver: and our successors may even dare to speculate on the changes that converted a crust of bread, or a bottle of wine, in the brain of Swift, Moliére, or Shakespeare, into the conception of the gentle Glumdalclitch, the rascally Sganarelle, or the immortal Falstaff. At present such thoughts would be justly regarded as the dreams of a lunatic, and I must crave your indulgence for having mentioned them. The history of science is, however, filled with such dreams; some never realized; others converted by time into realities so commonplace, that the genius of their originators is habitually forgotten or underrated.

During childhood and youth, the food that we eat is used for the double purpose of building up the tissues of the bones, muscles, brain, and other organs of the body; and of supplying the force necessary for Work done, whether mechanical or intellectual. In adult life, the first use of food almost disappears, for the bones, muscles, brain, and other organs, have already reached their full developement, and act simply as the media of communication between the Food received and the Work developed by it.

Let us take, as illustrations, the muscles and brain, regarded as the organs by means of which Mechanical and Intellectual Work is done. These organs resemble the piston, beam, and fly-wheel of the Steam Engine, and, like them, only transmit or store up the force communicated by the steam in one case, and by the products of the food conveyed by the blood in the other case. The mechanical work done by the steam engine must be measured by the loss of heat experienced by the steam in passing from the boiler, through the cylinder, to the condenser; and not by the loss of substance undergone by the several parts of the machinery on which it acts. In like manner, the mechanical or intellectual work done by the food we eat is to be measured, not by the change of substance of the muscles or brain employed as the agents of that work, but simply by the changes in the blood that supplies these organs—that is to say, undergone by the Food used, in its passage through the various tissues of the body, before it is finally discharged in the form of water, carbonic acid, or urea.

The Divine Architect has so framed the animal machine, that moves and thinks, that the same blood, which by its chemical changes produces movement and thought, also repairs the necessary waste of the muscles and brain, by means of which movement and thought are possible; just as if the steam that works an engine were able, without the aid of the engineer, to repair the wear and tear of its friction and waste spontaneously; but no greater mistake is possible in Physiology, than to suppose that the products of the changes in the blood, by which Mechanical or Intellectual Work is done, are themselves merely the result of the waste of the organs, whether muscles or brain, on the exercise of which that Work depends.*

The ancients, who derived all their knowledge from observation, and not from experiment, were well aware of the double duty imposed upon food in early life—of producing both the secular and the periodic variations of the body; or, in other words, of promoting growth, and of developing work.

* The very skill with which provision is made for the repair of the waste of the organ used as the instrument of Work may mislead the observer into supposing that the work itself may be measured by the waste of its instrument. Thus, it has been shown by Mr. A. Macalister, of Dublin, that the heart, which has imposed upon it the necessity of working day and night without ceasing, during life, is furnished with double the usual supply of blood through the coronary arteries, which are injected twice for every single beat of the heart. If, indeed, it were possible to assume that all muscles wasted equally for equal quantities of work, and also to measure separately the products of that waste, we might then assume the waste of the organ as the measure of its work. Neither of these assumptions, however, can be admitted, for it can be shown that different muscles act under different conditions, more or less advantageously, so that equal wastes would represent unequal works; and also, it is impossible to separate in practice the products of waste of muscles from those of the general changes of the blood.

Their practical knowledge is summed up by Hippocrates in the aphorism—

"Old men bear want of food best; next those that are adults; youths bear it least, more especially children; and, of these, the most lively are the least capable of enduring it."*

The food consumed in twenty-four hours, including air and water, undergoes a series of changes of a chemical character before leaving the body, in the form of one or other of its excretions. Some of these changes develope force, and others expend force, but the algebraic sum of all the gains and losses of force represents the quantity available for work. This work must be expended as follows:—

1. The Work of growth (*secular*).
2. The Work of maintaining Heat (*periodic*).
3. Mechanical Work (*periodic*).
4. Vital Work (*periodic*).

During childhood and youth the work of growth is positive, for a certain proportion of the food used is employed in building up the tissues of the body instead of being expended in actual work; it is, in fact, "stored up" in the body, as *vis viva* is stored up by the fly-wheel of machinery, and constitutes a reservoir of force that may be called upon at an emergency requiring sudden expenditure of force, as in case of illness; or to supply the gradual wasting of old age. In adult life, and in old age, the work of growth ceases completely, except so far as is necessary to repair, from day to day, the small wastes of the organs employed in Work; so that nearly the whole of the food employed is expended on the periodic work of the body. Hence we can readily see the reason for the aphorism, which asserts that food is more necessary for the young than

* *Γέροντες εὐφορώτατα νηστείην φέρουσι, δεύτερον οἱ καθεστηκότες ἥκιστα μειράκια, πάντων δὲ μάλιστα παιδία, τουτέων δὲ αὐτέων ἃ ἂν τύχῃ αὐτὰ ἑωυτῶν προθυμότερα ἐόντα.*—Aph. I. 13.

for the old, and more required by those of a lively disposition, either of mind or body, than by others.

Hippocratic Doctrine of Innate Heat.

Hippocrates was well aware of the connexion between food and animal heat, although he erroneously regarded the animal heat as an innate property of the body that caused an appetite for food, instead of being itself produced by food; if we transpose his cause and effect, *mutatis mutandis*, all his maxims as to animal heat are true. Thus, he says—

"Growing animals possess most innate heat, hence they require most food; . . . . . but the old have least heat, and therefore require the least fuel."*

"The cavities of the body are naturally warmest in Winter and Spring; . . . . in these seasons therefore most food must be given; and since there is more innate heat, more nourishment is required; as may be seen in youths and athletes."†

These maxims, when translated into modern language, express the well-known fact, that the chemical changes of food that take place in the body produce animal heat, and that the necessity for food to supply mechanical work is greatest with the young and active, while the necessity for the production of animal heat is greatest in the cold seasons of the year. The direct connexion of food with mechanical work is expressed in the following maxims:—

"There should be no labour when there is hunger"‡—

* Τὰ αὐξανόμενα πλεῖστον ἔχει τὸ ἔμφυτον θερμὸν, πλείστης οὖν δεῖται τροφῆς, ... γέρουσι δὲ ὀλίγον τὸ θερμὸν, διὰ τοῦτο ἄρα ὀλίγον ὑπεκκαυμάτων δέονται.—Aph. I. 14.

† Αἱ κοιλίαι χειμῶνος καὶ ἦρος θερμόταται φύσει. . . . . ἐν ταύτῃσιν οὖν τῇσιν ὥρῃσι καὶ τὰ προσάρματα πλείω δοτέον. Καὶ γὰρ τὸ ἔμφυτον θερμὸν πλεῖστον ἔχει, τροφῆς οὖν πλείονος δέονται. σημεῖον αἱ ἡλικίαι καὶ οἱ ἀθληταί.—Aph. I. 15.

‡ Ὅκου λιμὸς, οὐ δεῖ πονέειν.—Aph. II. 16.

and its converse,

"Let labour precede meals."*

On principles such as those just given, the training of the athletes was conducted; and they were compelled to undergo a regular course, commencing with blood-letting, and active purgation,† and consisting of systematic muscular exercise suited to the nature of the contest intended, accompanied by a dietary, of which the chief ingredients consisted of biscuits and pigs' kidneys, washed down by a minimum of water. It is, truly, not much to be wondered at, that those who survived the training were formidable in the boxing ring or racecourse.

The relation of animal heat to respiration is referred to by Hippocrates, in a remarkable maxim.

"Those persons have the loudest voices, who have most [innate] heat, for they inspire the largest quantities of the cold air; and the product of two great quantities must be itself great."‡

Galen believed the heart to be the centre of "innate heat," but he was well aware that increase or diminution of respiration caused increase or diminution of heat, and was intimately connected with it. Thus he says:

"Since, therefore, the heart is, as it were, the hearth and fountain of the innate heat, with which the animal is pervaded," &c.§

"The necessity for respiration is the greatest and most imperious guard of the innate heat."‖

* Πόνοι σιτίων ἡγείσθωσαν.—Epid. VI. Sect. iv. 28.

† Ἕλκεα ἐκφύουσιν ἣν ἀκάθαρτος ἐὼν πονήσῃ.—Epid. VI. Sect. v. 32.

‡ Οἷσι πλεῖστον τὸ θερμὸν, μεγαλοφωνότατοι, καὶ γὰρ ψυχρὸς ἀὴρ πλεῖστος. δύο δὲ μεγάλων μεγάλα καὶ τὰ ἔκγονα γίνεται.—Epid. VI. Sect. iv. 22.

§ ἐπεὶ τοίνυν ἡ καρδία τῆς ἐμφύτου θερμασίας, ᾗ διοικεῖται τὸ ζῶον, οἷον ἑστία τέ τίς ἐστι καὶ πηγὴ κ. τ. λ.—De usu partium, Lib. vi. ch. 7.

‖ ἡ χρεία τῆς ἀναπνοῆς ἡ μεγίστη μὲν καὶ κυριωτάτη φυλακὴ τῆς ἐμφύτου θερμασίας ἐστιν.—De diff. Resp. Lib. i. ch. 4.

"Those persons in whom the innate heat has been much cooled, breathe but little and slowly."*

Lavoisier's Theory of Animal Heat.

The doctrine of "innate" heat, taught by Hippocrates and Galen, ruled in Medicine for 1500 years after Galen's death; until it received its death blow from the genius of Lavoisier, who demonstrated in his celebrated memoir read before the French Academy of Sciences in 1783, that the source of animal heat is to be found in the combustion of the carbon of the body by the oxygen of the air received into the lungs by respiration. Lavoisier's experiments were repeated and confirmed in 1822 by Dulong and Despretz; and have formed the starting point for all modern investigations on the relation of food to work. As already stated, the work done by food in the body may be divided into

1. The Work of Growth.
2. The Work of Animal Heat.
3. Mechanical Work.
4. Vital Work.

Lavoisier arranged his experiments so as to exclude almost all the foregoing kinds of work, except that of animal heat. A Guinea-pig was placed under a bell glass inverted over a surface of mercury, and a current of fresh air was allowed to circulate through the apparatus, being passed at its final exit through tubes containing caustic potash, which arrested the carbonic acid produced by the animal. In this manner it was easy to ascertain the carbonic acid excreted, by the increase in weight of the tubes of caustic potash during the experiment.

Lavoisier found that his Guinea-pig, in ten hours, burned,

* ὥσπερ καὶ ὅταν μικρὸν εἰσπνέωσι καὶ βραδέως,· οἷς ἱκανῶς ἔψυκται τὸ ἔμφυτον θερμόν.—De diff. Resp. Lib. i. ch. 20.

on the average, 3.333 grms. of carbon; and this quantity of carbon he estimated from other experiments as capable of melting 326.75 grms. of ice at the freezing temperature. The same Guinea-pig was then placed in an ice calorimeter, and left in it for ten hours, during which time the heat of its body was found to have melted 402.27 grms. of ice at the freezing temperature.

If we use, instead of the coefficient of combustion of carbon employed by Lavoisier, that now generally adopted from the experiments of Favre and Silbermann, the quantity of melted ice represented by $3\frac{1}{3}$ grms. of carbon would become 364.78 grms., instead of 326.75 grms. We are, therefore, entitled to say that the heat of combustion of expired carbon determined by Lavoisier is equal to

$$\frac{36478}{402.27} = 90.68 \text{ per cent.}$$

of the animal heat developed, which is regarded as 100 parts.

Two years later, in 1785, Lavoisier laid before the Royal Society of Medicine of Paris, an account of further experiments, also conducted on the breathing of Guinea-pigs, by which he showed, that of 100 parts of oxygen absorbed by those animals, 81 only reappeared in the form of carbonic acid, and 19 parts disappeared altogether. Lavoisier considered that these 19 parts of oxygen were employed in the body in the combustion of hydrogen, the product of such combustion being water.

If we use Lavoisier's data just given, and the known atomic weights of carbon, oxygen, and hydrogen, we shall have, for 81 parts of oxygen in the form of carbonic acid, and 19 parts of oxygen in the form of water, the following quantities of carbon and hydrogen consumed by the respiration of his Guinea-pig in the same time:—

$$\text{Carbon} = \frac{6 \times 81}{16} \quad \text{Hydrogen} = \frac{19}{8}$$

Multiplying these numbers by the Heat Coefficients of Favre and Silbermann, we find—

$$\text{Heat produced by Carbon} = \frac{6 \times 81}{16} \times 8080$$

$$\text{Heat produced by Hydrogen} = \frac{19}{8} \times 34462$$

It has been already shown that the heat developed by the combustion of carbon in Lavoisier's experiment amounted to 90.68 per cent. of the heat emitted by the animal; hence the heat produced by the combustion of the hydrogen will amount to

$$90.68 \times \frac{19 \times 34462}{8} \times \frac{16}{6 \times 81 \times 8080} = 30.24$$

By adding together the heats due to the carbon and hydrogen, we find that Lavoisier's experiments, when fairly interpreted by the data of modern science, give the following results:—

Heat produced by the combustion of carbon and hydrogen, ...	120.92
Animal Heat, ...	100.00

Finally, in 1789, Lavoisier published further experiments, by which he showed conclusively that the consumption of oxygen by the body is notably increased by three causes—

1°. By a lowering of the external temperature.
2°. By the act of digestion.
3°. By muscular exercise.

The experiments of Lavoisier were repeated in 1822 by Dulong and Despretz, and their results, when corrected, like those of Lavoisier, by using the modern heat coefficients of carbon and hydrogen, are as follows:—

The mean of Dulong's experiments on 16 animals and birds is 90.6 per cent. of the animal heat given out—the lowest number, 85.5, belonging to a kitten 60 days old; and the highest number, 99.4, belonging to a puppy 50 days old.

M. Despretz obtained an average of 92.3, from 16 mammals and birds; his highest number being 101.8, derived from an old female rabbit; and his lowest number being 84.2, derived from 4 owls.

The foregoing experiments left no doubt remaining in the minds of men of science as to the substantial truth of Lavoisier's doctrine of animal heat; and led immediately to a number of supplementary experiments, amongst the most remarkable of which were those of Regnault and Reiset.

Regnault directed his attention especially to the distribution of the oxygen absorbed by animals, between the carbon and hydrogen of their blood, or tissues, which had been laid down by Lavoisier in the proportion of 81 to 19. He found that the proportion was not a fixed one, but varied with the food in a very instructive manner.

The average of his experiments on 14 animals, including worms, lizards, and insects, as well as birds and mammals, was—

Oxygen combined with carbon, . .	81.7
Oxygen combined with hydrogen, .	19.3

a result nearly identical with that found by Lavoisier. The highest proportion of oxygen combined with hydrogen, occurred in the case of chickens fed on meat, and amounted to 32 per cent.; and the lowest proportion occurred in the case of rabbits fed on bread and oats, and amounted to 1 per cent. only.

Still more recent experiments, made with improved apparatus and methods by Pettenkofer and Voit, in Munich, show, like those of Regnault, that the proportion of the oxygen employed in forming carbonic acid, to the whole oxygen absorbed, varies with the food, ranging in the case of a large dog from 52.4 to 148.2, according as the animal was kept altogether without food, or fed upon a mixed diet of meat and sugar. These investigations have also shown that, under ordinary conditions, it is probable that a dog consumes nearly all the oxygen absorbed in the formation of carbonic acid.

Before leaving the subject of animal heat, it is worth while to estimate its amount in a manner that will bring it into comparison with ordinary mechanical work.

In Lavoisier's experiment with the Guinea-pig, 402.27 grms. of ice were melted in ten hours; from this fact we find, assuming the latent heat of ice at 142° F., and 772 as Joule's coefficient for converting British heat units into foot pounds,

Mechanical work equivalent to the daily animal heat of Lavoisier's Guinea-pig =

$$\frac{402.27 \times 24 \times 142 \times 15.432 \times 772}{7000 \times 10}$$
$$= 233310 \text{ ft. lbs.}$$

As the average weight of a Guinea-pig is 4 lbs., the preceding amount of work, representing animal heat, would be sufficient to raise the weight of the animal through a vertical height of

$$\frac{233310}{4 \times 5280} = 11.05 \text{ miles.}$$

Ranke has shown, by experiments made upon himself, under various conditions of food and fasting, by means of Pettenkofer and Voit's apparatus, that his daily excretion of carbonic acid varied from 660 grms. to 860 grms., showing a mean of 760 grms. His weight was 67 kilos., from which fact, and the assumption that an English mile is 1600 meters, we obtain, employing the constants already given, the height through which the combustion of 760 grms. of carbonic acid would raise the weight of 67 kilos. in 24 hours—

$$= \frac{760 \times 6 \times 8.080 \times 423}{22 \times 67 \times 1600} = 6.609 \text{ miles.}$$

The extreme values of the carbonic acid excreted, viz. 660 grms. and 860 grms. would correspond to the heights of 5.74 miles, and 7.48 miles respectively.

Dr. Edward Smith has estimated the daily excretion of carbon from the lungs, in the case of four persons, as follows:—

	Body Weight.	Carbon.
Mr. Moul, . . .	173 lbs.	6.735 oz.
Dr. E. Smith, . .	196 „	7.85 „
Prof. Frankland, .	136 „	5.60 „
Dr. Murie, . . .	133 „	6.54 „

In order to convert the preceding data into vertical miles through which the body weight is lifted, we must multiply the ounces of carbon by the following coefficient, and divide the product by the body weight.

$$\text{Coeff.} = \frac{8080 \times 9 \times 772}{16 \times 5 \times 5280} = 132.91$$

$$\log. (\text{coeff.}) = 2.1235473$$

We thus obtain, for the heights through which the carbon consumed would lift the observers—

Mr. Moul, ...	5.17 miles.
Dr. E. Smith, ...	5.32 „
Prof. Frankland, . .	5.47 „
Dr. Murie, ...	6.53 „

Pettenkofer and Voit succeeded in producing a range of carbonic acid excreted by a large dog, weighing 33.3 kilos., from 289.4 grms. to 840.4 grms; the minimum corresponding to the 10th day of fasting from solid food, and the maximum corresponding to a diet of 1800 grms. of meat, 350 grms. of fat, and 1410 grms. of water.

It may be easily shown by a calculation similar to the foregoing, that these excretions of carbonic acid correspond to the mechanical works of lifting the weight of the dog through vertical heights of 5.03 miles, and 14.62 miles respectively.

Combining together the preceding results, and expressing them all in the natural units of the weights of the animals lifted through a height, we find—

Work due to Animal Heat.

MAN.

1. Dr. Ranke (fasting), ...	5·74 miles.
2. Dr. Ranke (well fed), . .	7·48 ,,
3. Mr. Moul, ...	5·17 ,,
4. Dr. E. Smith, ...	5·32 ,,
5. Prof. Frankland, ...	5·47 ,,
6. Dr. Murie, ...	6.53 ,,
Mean, ...	**5.952** miles.

This result agrees very closely with the calculation already made from 760 grms. of carbonic acid, in the case of Dr. Ranke; viz. 6.609 miles.

Work due to Animal Heat.

ANIMALS.

1. Guinea-pig, ...	11.05 miles.
2. Dog (fasting), ...	5.03 ,,
3. Dog (overfed), ...	14.62 ,,
Mean, ...	**10.233** miles.

Source of Muscular Work.

As soon as it was satisfactorily established by Lavoisier and his successors that the natural combustion of carbon and hydrogen in the blood was sufficient, or somewhat more than sufficient, to account for the animal heat, it became a matter of great interest to physiologists to ascertain, if possible, how much of the work developed in the blood by chemical changes is employed in producing animal heat, how much in mechanical work, external and internal, and how much in vital or mental operations.

At the outset of this inquiry, it received a misdirection from the conjecture thrown out by Liebig, that the excretion of nitrogen (in the form of urea) gave necessarily the measure of the wear and tear of the muscular tissues themselves, which are composed of proteinic or nitrogenous compounds. This conjecture led to Liebig's celebrated classification of food into Heat-producing and Flesh-forming foods, which has been unhesitatingly received until lately, in this country, by physiologists and physicians. Before investigating the truth or falsehood of Liebig's theory, it is worth while to state the most recent results obtained as to the muscular work per day of which man is capable.

From numerous observations, of which some were made by myself on the daily labour of hodmen, paviours, navvies, and pedlars, I have obtained the following mean:—

Daily labour of Man = 353·75 ft. tons. = 109549 kil. met.

This quantity of work is the exact equivalent of the work done by a man of 150 lbs. weight in climbing through one mile of vertical height, and is, as I have already shown, about one-sixth part of the work expended in producing and maintaining animal heat.

I was led to believe, from investigations made to determine

the quantity of urea excreted in various diseases, that a certain minimum quantity, equivalent to 2 grs. per pound of body weight, was excreted quite independently of muscular exertion, and I proved that death was preceded in many chronic diseases by a fall in the urea excreted to 2 grs. per pound. These investigations were made chiefly on patients dying of advanced kidney disease, in which the excretion of albumen had nearly or altogether ceased, and on patients dying of phthisis.

Pettenkofer and Voit found that the excretion of urea in a dog reduced from 33.3 kilos. to 29 kilos. by 10 days' fast became 8.6 grms. And, since

29 kilos = 63.8 lbs.
8.6 grms. = 132.7 grs.
Excretion of urea = 2.08 grs. per lb. of body weight.

Ranke obtained a precisely similar result from observations made upon himself, after long fasting, continued for several days.

If these views be well founded, it is plain that part only of the urea excreted can be regarded as due to muscular exertion, for 2 grs. per lb. (or 300 grs. for a man weighing 150 lbs.) must be set aside as a constant due to vital work, independent of muscular work altogether. Hence it would follow, supposing the muscular exertion to be measured by the increased excretion of urea produced by it, that the urea will not increase as fast as the muscular exertion, but it ought to increase regularly, although at a slower rate. With a view to settle this important question, I devised the following observations upon myself in the month of July, 1866, which prove conclusively that an increase of muscular exertion, amounting to fourfold, is not accompanied by any corresponding increase in the excretion of nitrogen, in the form of urea.

I had previously ascertained by repeated experiments, extending from 1860 to 1865, that my excretion of urea (under ordinary conditions as to exercise, which never amounted to five miles per day), ranged from

465.09 grs. per day, to
537.47 grs. per day.

501.28 mean.

This quantity of urea I regarded as my natural physiological average, and it was so well established, that I thought I should obtain an important result by comparing it with the average found from several days of unusual muscular exertion. I accordingly walked for five consecutive days in the hilly districts of Wicklow, noting carefully the horizontal distance travelled each day, and the vertical height traversed up and down. The vertical heights were reduced to horizontal distances, on the assumptions (which are well founded) that 20 is the proper coefficient for converting one into the other, and that the work of descent is half the work of ascent.

During the five days of observation the work done, expressed in horizontal miles of walking, was as follows:—

First Day.

		Miles.
Miles walked, ...		11.4
Height ascended, ...	1800 ft. =	10.2
		21.6

Second Day.

Miles walked, ...		12.0
Height ascended, ...	2400 ft. =	13.7
		25.7

Third Day.

Miles walked, ...		11.6
Height ascended, ...	1400 ft. =	8.0
		19.6

Fourth Day.

Miles walked, ...		9.3
Height ascended, . . .	1400 ft. =	8.0
		17.3

Fifth Day.

Miles walked, ...		10.4
Height ascended, ...	1600 ft. =	9.1
		19.5

From the preceding statement it follows that the average work done each day was 20.74 miles of horizontal walking—the result of which upon the urea excreted was to be compared with the result already mentioned, as a physiological constant determined under circumstances in which the daily muscular work never exceeded 5 miles of horizontal walking.

In order to determine the urea, I collected each day all the urine passed, and kept one-fifth part of it; and at the close of the fifth day examined the mixture formed from the five days' urine. It was found to contain 501.16 grs. of urea per day—a result practically identical with the physiological quantity previously found by me under totally different conditions, viz. 501.28 grs. I was much surprised at this result, for I had previously believed in the theory laid down by Liebig, which attributed the excretion of urea to the disintegration of muscular tissue.

It might be objected to the preceding reasoning, that the combustion of proteinic compounds represented by 501.28 grs. of urea excreted is actually sufficient to produce the mechanical force necessary to maintain the muscular exertion of walking 20 or 21 miles per day.

1°. The urea excreted bears to the proteine consumed the proportion of 24 to 79; as appears from their chemical compositions—viz.,

Urea, . .	$C_2\ H_4\ N_2\ O_2$ . .	60
Proteine, .	$C_{36}\ H_{27}\ N_4\ O_{12}$ . .	395

2°. In 100 parts of proteine there are 53.7 parts of carbon, and 7 parts of hydrogen; the total heat due to the combustion of 1 grm. of proteine is, therefore,

	Heat Units.
0.537 grm. of Carbon, ...	4.3389
0.070 grm. of Hydrogen, ...	2.4123
	6.7512

This number, 6.7512, represents the maximum quantity of heat units* that could be produced by the combustion of 1 grm. of proteine; but the term depending on hydrogen in it should be reduced to $\frac{5}{9}$ths of its amount, in consequence of the hydrogen already combined with oxygen in the proteine. Hence we find—

Combustion of 1 grm. of Proteine.

Carbon, ...	4.3389 heat units.
Hydrogen, ...	1.3402 ,,
	5.6791

3°. In 100 parts of urea there are 20 parts of carbon, and $6\frac{2}{3}$

* Heat unit = 1 kilog. of water raised 1° C.

parts of hydrogen; the total heat, therefore, due to the combustion of 1 grm. of urea is

0.20 grm. Carbon, ...	1.6160
0.067 grm. Hydrogen, ...	2.3089
	3.9249

The term depending on hydrogen, in this result, should be reduced to $\frac{1}{2}$, in consequence of the hydrogen already combined with oxygen in the urea.

Hence we find—

Combustion of 1 grm. of Urea.

Carbon, ...	1.6160
Hydrogen, ...	1.1544
	2.7704

4°. From the three preceding statements it is easy to see that, for every gramme of proteine consumed, 0.8416 heat units are contained in the urea excreted; so that

The Digestion of 1 grm. of proteine gives out 4.8375 *heat-units.*

It is easy to see that 501.28 grs. of urea excreted correspond to 1650 grs. of proteine in the food, or to 106.92 grms.; and the total work due to the digestion of this quantity of food may be found by multiplying it by the "*Digestion coefficient*" already found, and by 423, which is Joule's coefficient for converting heat-units into kilogrammeters. Hence we have—

Work due to production of 501.28 *grs. of urea*

$$= 106.92 \times 4.8375 \times 423$$
$$= 218786 \text{ k. m.}$$
$$= 704 \text{ ft. tons.}$$

This amount of theoretical work produced by nitrogenous food is double the work actually done during the walking excursion.

The average work was 20.74 miles horizontal per day, which may be considered as the exact equivalent of lifting my weight (knapsack and clothes included = 150 lbs.) through one mile of vertical height. Hence the work actually done by me was

$$\frac{150 \times 5280}{2240} = 354 \text{ ft. tons.}$$

This amount of muscular work accounted for almost exactly half the whole theoretical work supplied by the food that goes to form urea, viz. 704 ft. tons.; but it has been already shown that 2 grs. of urea per pound of body weight is required to maintain the vital work, including circulation and respiration; this would give (since I weighed 128 lbs.) 256 grs. of urea, required for vital work; or almost exactly half of the 501.16 grs. excreted; so that one-half of the available work might be considered as expended on vital work, and the other half as expended on external muscular work. This supposition, however, requires us to believe that the muscles act without loss by friction. This is not admissible, for I have elsewhere endeavoured to show that there is a loss in the force applied by the muscles of various animals, in consequence of the friction of their tendons amounting, on the average, in man to 35 per cent., and in the mastiff to 41 per cent.

Hence it may be regarded as certain that the available force represented by 501 grs. of urea is not sufficient to account fully, both for vital work and for the external mechanical work expended by me during the experiments just described.

The foregoing observations and calculations were made in

the month of July, 1866, but I did not then publish them, as I found afterwards that I had been anticipated by Dr. Fick and Dr. Wislicenus of Zurich, in a paper published in June in the Philosophical Magazine, on the urea excreted during an ascent of the Faulhorn. Professor Frankland, in a paper published in the same journal in September, 1866, corrected some erroneous reasoning that found its way into Fick's and Wislicenus' paper, and further supplied, from direct experiment, the *Digestion coefficient* of proteine, which had been obtained by me from calculation. The actual value of this important constant was found by him to be—

Actual value of digestion coefficient of Proteine,	4.3155
Calculated value, ...	4.8375

My only object in now publishing an account of the independent experiment and calculation made by myself, is to confirm the certainty of the important fact first proved experimentally by Fick and Wislicenus, that the force due to the urea excreted in a given time is not sufficient to provide the actual work that may be done by the muscles in the same time.

Liebig and his followers, misled by a preconception of the simplicity of nature, assigned to nitrogenous food the duty of providing the force necessary for the production of muscular work, by supplying the waste of muscular tissue; while they supposed the farinaceous and fatty foods to provide the amount of animal heat required by the body.

The opponents of Liebig have fallen into the opposite error, and deny that nitrogenous food contributes any portion of the force employed in muscular work.

The truth, as is usual, lies between the two extreme hypotheses, and we are now compelled to admit that a given developement of force, expressed in animal heat, muscular work, and mental exertion, may be the effect of several, perhaps many, supposable supplies of digested food, farinaceous, saccharine, fatty, and albuminous.

Just as a given algebraical function may be equated to a given constant, by the use of a certain definite number of values of its variable quantity, so may a given effect of work in the animal body be produced by certain definite, though very different, combinations of various kinds of food, the digestion of which follows each its own law, and developes its own amount of force. The number of roots in our equation of life increases the difficulty of solving it, but by no means permits the acceptance of the lazy assumption that it is altogether insoluble, or reduces a sagacious guess to the level of the prophecy of a quack.

Lavoisier supposed in his earlier investigations that animal heat was developed by the combustion of carbon and hydrogen in the lungs; just as in earlier times it was supposed to be produced spontaneously in the heart, which was imagined to be so hot as even to burn the hand that should imprudently venture to touch it.

In like manner, Liebig and his followers supposed the muscular work to be developed in the substance itself of the muscles that were its instruments.

Both of these doctrines are now justly repudiated by physiologists, and the view, proposed in 1845, by Dr. Mayer of Heilbronn, and recently developed with much ability, by Mr. C. W. Heaton, of Charing Cross Hospital, in the Philosophical Magazine for May, 1867, that the blood itself is the seat of all the chemical changes that develope force in the body, has gained favour among physiological chemists, and also met with acceptance among practical clinical observers.

Thus the human mind revolves in cycles, and the physicians of the nineteenth century are preparing to sit at the feet of Moses, and learn that the blood of an animal really constitutes its life; while South African theologians are disposed to reject his authority, because he happened to confound a Rodent with a Ruminant.*

Whatever be the kind of food employed, its effect in the production of force must be ultimately measured by the quantities of carbonic acid and water produced by its combustion, and there is no more convenient measure of the production both of carbonic acid and water than urea, so far as it goes. I shall prove shortly that every four grains of urea excreted represent five tons lifted through one foot; and I have shown by the preceding investigation that the work represented by urea is not sufficient to account for vital and external work, much less for animal heat. The investigations of Dr. Edward Smith, on the excretion of carbonic acid, enable us to show that the carbonic acid alone is sufficient to account for both vital and external work, and also for the production of animal heat. This may be proved as follows:—

Dr. Smith has given results, from which may be deduced the quantities of carbonic acid excreted per minute, during the four following conditions:—

1. Lying in the horizontal position, and nearly asleep.
2. Fasting, and in sitting posture.
3. Walking at two miles per hour.
4. Walking at three miles per hour.
5. Working on the Treadmill, ascending at the rate of 28.65 feet per minute.

	Carbonic acid per min.
1. Sleep and rest, ...	5.522 grs.
2. Sitting, ...	7.440 „
3. Walking at two miles per hour, .	18.100 „
4. Walking at three miles per hour, .	25.830 „
5. Treadmill, ...	44.973 „

* No reasonable person can fail to perceive the ignorance of the great Lawgiver, who will apply to him the test first proposed by Swift for Homer; Moses, like the author of the Iliad, was profoundly unacquainted with the discipline and doctrines of the Church of England.

The foregoing quantities of carbonic acid per minute may be converted into vertical miles per hour for the body weight, by multiplying them by the following coefficient:*—

$$\frac{60 \times 6 \times 8080 \times 9 \times 772}{22 \times 196 \times 5280 \times 5 \times 7000} = 0.025263$$

$$\log. = \bar{2}.40420$$

Performing this calculation we find—

	Carbonic Acid.	Body weight lifted through miles.
1.	5.522 grs.	0.1400 mile.
2.	7.440 ,,	0.1887 ,,
3.	18.100 ,,	0.4591 ,,
4.	25.830 ,,	0.6551 ,,
5.	44.973 ,,	1.1406 ,,

It is easy to calculate that the external work done in the cases 3, 4, 5, was as follows:—

	External Work.
No. 3. Walking two miles per hour, .	0.1000 mile.
No. 4. Walking three miles per hour, .	0.1500 ,,
No. 5. Treadmill, ...	0.3256 ,,

Subtracting these amounts of work from the applied work, due to the production of carbonic acid, we find, as the quantities left for Vital Work, including circulation and respiration, and for the production of Animal Heat, per hour:

	Vital Work and Animal Heat.
No. 3.	0.3591 mile.
No. 4.	0.5051 ,,
No. 5.	0.8150 ,,

As I have already shown the work due to animal heat per day to be 6 miles; it follows that the work of animal heat per hour is 0.2500 mile.

* Dr. Edward Smith's weight was 196 lbs.

Deducting this amount from the foregoing, we find for the Vital Work done, under the three different conditions—

	Vital Work.
No. 3. Walking at two miles per hour,	0.1091 mile.
No. 4. Walking at three miles per hour,	0.2551 „
No. 5. Treadmill work, ...	0.5650 „

This result proves, in a striking manner, the great disadvantage under which an increased amount of muscular work is done, in a given time; and it is quite in accordance with other results obtained by me from totally different experiments.

No two classes of animals can well differ more from each other than the Cats and Ruminants, one of which is intended by nature to eat the other. They differ in all respects as to food, the Cats requiring a supply of fresh meat and blood for their health, and the Ruminants being exclusively vegetable feeders; yet in both classes we find a great developement of muscular power, and of rapid action of muscles, qualities alike necessary to the pursuer and to the pursued. There can be no doubt that muscular work is developed in the Cats from the combustion of flesh, and in the Ruminants, mainly, if not exclusively, from farinaceous food. It is, however, worthy of remark, that the muscular qualities developed by the two kinds of food, differ considerably from each other. The hunted deer will outrun the leopard in a fair and open chase, because the work supplied to its muscles by the vegetable food is capable of being given out continuously for a long period of time; but in a sudden rush at a near distance, the leopard will infallibly overtake the deer, because its flesh food stores up in the blood a reserve of force capable of being given out instantaneously in the form of exceedingly rapid muscular action.

In conformity with this principle, we find among ourselves an instinctive preference given to farinaceous and fatty foods, or to nitrogenous foods, according as our occupations require a steady, long-continued, slow labour, or the exercise of sudden bursts of muscular labour continued for short periods. Thus Chamois hunters setting out for several days' chase provide themselves with bacon fat and sugar; the Lancashire labourers use flour and fat, in the form of apple dumplings; while the Red Indian of North America almost transforms himself into a carnivore, by the exclusive use of flesh food; he sleeps as long, and can fast as long as the Puma or Jaguar, and possesses stored up in his blood a reserve of force which enables him, like a cat, to hold his muscles for hours in a rigid posture, or to spring upon his prey, like a leopard leaping from a tree upon the back of an antelope.

If the preceding view of the muscular qualities developed by the two kinds of food be correct, important inferences suggest themselves as to the food that should be employed in relation to several kinds of work. Of these inferences, I shall select two examples:—

1. The nurses of one of our Dublin Hospitals were formerly fed chiefly upon flesh food and beer, a diet that seemed well suited to their work in ordinary times, which was occasionally severe, but relieved by frequent intervals of complete rest. Upon the occasion of an epidemic of cholera, when the hospital duties of the nurses became more constant, although on the whole not more laborious, they voluntarily asked for bacon fat and milk, as a change of diet from the flesh meat and beer; this change was effected on two days in each week with the best results as to the health of the nurses, and as to their power of discharging the new kind of labour imposed upon them.

2. I have been informed, on competent authority, that the health of the Cornish miners breaks down ultimately, from failure of the action of the heart and its consequences, and not from the affection of the lungs called "miner's phthisis." The labour of the miner is peculiar, and his food appears to me

badly suited to meet its requirements. At the close of a hard day's toil, the weary miner has to climb by vertical ladders through a height of 100 to 200 fathoms before he can reach his cottage, where he naturally looks for his food and sleep. This climbing of the ladders is performed hastily, almost as a gymnastic feat, and throws a heavy strain (amounting from one-eighth to one-quarter of the whole day's work) upon the muscles of the tired miner, during the half hour or hour that concludes his daily toil. A flesh-fed man (as a Red Indian) would run up the ladders like a cat, using the stores of force already in reserve in his blood; but the Cornish miner, who is fed chiefly upon dough and fat, finds himself greatly distressed by the climbing of the ladders—more so indeed than by the slower labour of quarrying in the mine. His heart, over-stimulated by the rapid exertion of muscular work, beats more and more quickly in its efforts to oxidate the blood in the lungs, and so supply the force required. Local congestion of the lung itself frequently follows, and lays the foundation for the affection, so graphically, though sadly, described by the miner at 40 years of age, who tells you that "his other works are very good, but that he is beginning to leak in the valves."

Were I a Cornish miner, and able to afford the luxury, I should train myself for the "ladder feat" by dining on half a pound of rare beefsteak and a glass of ale, from one to two hours before commencing the ascent.

The excretion of nitrogen by the Cats and Ruminants is very different, as might be expected from their food. I have ascertained that the urea discharged by a Bengal Tiger and a Sheep, daily, is as follows:—

Bengal Tiger, ...	4375	grs. of urea.
Sheep, ...	256	„

It is worthy of remark, and serves to throw light on the meaning of the excretion of nitrogen from the body, that causes but slightly connected with muscular exertion in the Ruminants increase amazingly the excretion of urea. Thus I have found the following excretion of urea from a Ram during the rutting season:

Ram (rutting season), . . . 1493 grs. of urea.

This amounts to a *sixfold* increase of urea, which cannot possibly be accounted for by the food consumed at the time, but requires us to assume a certain storing up of force, represented by nitrogenous compounds, which has been going on for a considerable period previous to the rutting season. A similar and equally remarkable storing up of phosphates and carbonates takes place, previous to the rutting season, in the Ruminants that shed their horns, which in the *Cervus Megaceros* often weigh 90 lbs.

These remarkable phenomena remind us of the maxim of the wise Hippocrates, who recommends moderation in the use of the gifts of the Golden Venus, as well as in those of Ceres and Bacchus—

πόνοι, σιτία, ποτὰ, ὕπνος, ἀφροδίσια μετρια,

with which may be compared its converse in the Latin proverb—

Sine Cerere et Baccho, friget Venus:

or, as the old proverb says;

When the wolf comes in at the door, love flies out
at the window.

Application of Theory to Diseased Conditions of Body.

The relation of food to work, complicated enough in health, becomes more so in disease, and the problem to be solved by rational theory becomes still more difficult. I cannot attempt even to sketch an outline of this part of my subject considered in general, but shall content myself with asking your attention to three remarkable examples of disease which illustrate the principles I have attempted to lay down.

These diseases are—

A. Typhus Fever.
B. Cholera Asiatica.
C. Diabetes mellitus.

A. *Typhus Fever.*—In Typhus fever a prominent symptom is the remarkable elevation of temperature, accompanied by an increased excretion of urea and carbonic acid, by the kidneys and lungs, indicating (as no food is taken) an increased morbid metamorphosis of the blood and tissues. The temperature commonly rises to 104° F., representing an increase of upwards of 5° F. above the normal temperature.

If we knew the cause of this increase of temperature, or rather of the increased metamorphosis of which it is the sign, we should know the cause of *Typhus* fever, and learn to combat the disease on rational grounds. At present the cause is unknown, and therefore the physician is forced to treat the symptoms as they appear, instead of attacking the cause of the disease. Let us examine for a moment the terrible significance of the symptoms.

Your patient lies for nine or ten days, supine, fasting, subdelirious; the picture of weakness and helplessness; and yet this unhappy sufferer actually performs, day by day, an amount of work that might well be envied by the strongest labourer in our land.

The natural temperature of the interior of the body is 100° F., while the temperature of the corresponding parts in Typhus fever is at least 105° F. This seems at first sight a small increase—only 5 per cent. of the whole; but it is in reality $2\frac{1}{2}$ times as great as it appears, and actually amounts to $12\frac{1}{2}$ p. c., or one-eighth part of the total animal heat. For the total quantity of heat given out by the heated body is proportional (from Newton's law of cooling) to the elevation of its temperature above the temperature of equilibrium, towards which it tends. If we suppose this equilibrium temperature to be 60° F., then the quantities of animal heat given out in Typhus fever and in health will be in proportion of 45 to 40, showing that the animal heat of Typhus exceeds that of health by one-eighth of its amount.

We have already seen that the work due to Animal Heat would lift the body through a vertical height of 6 miles per day; and it thus appears that an additional amount of work, equivalent to the body lifted through nearly one mile per day, is spent in maintaining its temperature at Fever Heat.

If you could place your fever patient at the bottom of a mine, twice the depth of the deepest mine in the Duchy of Cornwall, and compel the wretched sufferer to climb its ladders into open air, you would subject him to less torture, from muscular exertion, than that which he undergoes at the hand of nature, as he lies before you, helpless, tossing, and delirious, on his fever couch.

The treatment of this formidable disease in former times consisted of purging, vomiting, and bleeding the patient, with the view of eliminating an imaginary poison, and so helping nature to terminate the disease.*

In modern times, thank God, the physician either does not interfere at all; or adopts the rational process of retarding the disintegration of the tissues consumed to supply the fever heat, by furnishing in their stead, fuel, in the form of wine and beef tea, sufficient to maintain the increase of temperature imperiously required.† This practice may be justly considered

* *Νούσων φύσιες ἰητροί.*—Epid. vi. Sect. v. 1.

† It is not intended by this to assert that a high temperature, 104° to 108° F., must be maintained, in order that the disease may terminate favourably, for the very contrary is the fact. The blood, in Typhus, as in other pyrexies, is a fluid

rational, because the condition of the circulation admits of its application, and it is considered good, because it has been rewarded with success, in the hands of the skilful clinical physician. In concluding this sketch of the prominent symptom of Typhus fever, and as an illustration of the eagerness with which every possible combustible in the body is made use of, I may mention, on the high authority of Dr. Stokes, of Dublin, that the very urea excreted by the kidneys is not permitted to leave the body without first paying its tax to fever, by being burned into carbonate of ammonia, thus rendering the urine of an advanced case of bad Typhus fever eminently alkaline.

B. *Asiatic Cholera.*—This remarkable disease presents, as every one knows, three distinct stages, viz.,

possessed of greater oxidising power than it has in health; in consequence of this, an increased metamorphose of tissues takes place, accompanied of course by an elevation of temperature, which measures precisely the oxidising power of the blood, and the risk to life in Typhus is directly proportional to the rise in temperature. The indications of the sphygmograph are similar to those of the thermometer, a "*full dicrotic*" pulse corresponding to a temperature of 103° F., and the pulse of "*death agony*," with the heart's first sound gone, corresponding to a temperature of 109° F. There is no case on record of recovery from a condition marked by such a pulse and temperature.

The effects of alcohol, administered in fever, when the temperature does not exceed 105° F., are twofold—immediate and secondary. The immediate effect is to supply a hydrocarbon to the blood, which is decomposed by it in preference to the body tissues. The secondary effect of alcohol is to change the blood itself, which thus loses its oxidising qualities; in consequence of which the temperature falls, the hyperdicrotic character of the pulse disappears, and the destructive metamorphose of the tissues becomes lessened. The statement here given of the effects of alcohol given in Typhus, to the exact amount required by the condition of the blood, in narcotic doses, is borne out by clinical observation, and is independent of any theory as to the cause of Typhus.

It is not at all improbable that the theory of contagious disease, that each such disease owes its existence to a special living organism, and not to an organic poison, may ultimately prove to be correct.

1. The premonitory stage of diarrhœa.
2. The stage of collapse.
3. The stage of consecutive fever.

The stage of collapse exhibits the following symptoms :—vomiting or purging; muscular cramps; suppression of bile and urine; lowering of body temperature to 95° F.; extreme prostration of strength; extremities pulseless; and face Hippocratic.

When death occurs during collapse, the following symptoms are usually found, on careful examination of the corpse. The temperature rises to 103° F.; the muscles give out their characteristic susurrus CCC, and exhibit spontaneous movements; the whole train of symptoms producing the effect of a ghastly attempt at resurrection.*

In this disease we have phenomena respecting animal heat, the very reverse of those found in Typhus fever; the body performing one vertical mile short of its daily work, instead of one mile in excess. The prostration of strength resulting from this deficient combustion is so great, that death is often caused by bringing the patient to hospital in a cab instead of upon a stretcher, by his walking up a dozen steps into his ward, and sometimes even fatal results have followed a sudden effort to sit up in bed to vomit.

The rise of temperature after death, and the continuance of muscular susurrus and motion, tend to prove that the impeded circulation which is the prominent symptom in Cholera collapse, is due to constriction (probably vasomotor nervous) of the capillaries—in consequence of which the muscles are deprived of their supply of freshly oxidised blood, the result

* It is startling, on making a post-mortem examination of a cholera patient alone, and by candle light, to witness, on the first free incision of the scalpel, the hand of the corpse rise slowly from its side and placed quietly across its breast.

of which is necessarily contraction, and cramp, which produces the excessive agony that characterizes this disease.

All authorities on Cholera, whether their object be to "impede" or to "assist" Nature, are agreed that medicines, whether astringent or purgative, are not only useless, but dangerous in the stage of collapse.

It is useless to give alcoholic fuel to restore the loss of animal heat, for there is no circulation to cause the oxidation of the hydrocarbons.

It is equally useless and more dangerous to give opium, to check the remaining purging that exists; for if vomiting have ceased, your acetate of lead and opium pills lie, as if in the stomach of a corpse, and at the termination of collapse, your patient enters upon the consecutive fever, with perhaps a dozen grains of opium in his stomach, placed there like an explosive shell by your ill-timed zeal, and rapidly passes into a comatose condition, from which he never for a moment rallies. His death is always accredited by the Registrar to cholera morbus, and not to opium.

Purgative sand emetics* in Cholera collapse effect the same object as opium, but with greater rapidity. In the stage of blue collapse, the chances of life and death are almost exactly equal, and the slightest additional loss of force turns the wavering beam on the side of death. The effects of a brisk purgative or emetic (if they act) upon a patient, unable to climb a dozen steps, or sit up for a quarter of an hour, without fatal syncope, may be easily imagined; and the use of them cannot be justified by any arguments borrowed from right reason.

A remarkable though transient improvement takes place in Cholera collapse by the injection of warm water (brought to the specific gravity of serum by the addition of mineral salts) into the veins or bowels; the patient loses the cramps, feels

* When mustard is used, its conservative effects as a stimulant sometimes counteract its destructive effects as an emetic.

that he is about to recover, speaks to his friends, and often transacts whatever business is necessary; but speedily falls back into collapse. The improvement in his condition is altogether due to the temperature of the fluid injected, which supplies for a brief period the deficient animal heat, permits a partial oxidation of the blood, restores the capillary circulation in the muscles, and so destroys their cramp ; and by supplying the deficient work required, removes for the moment the fatal prostration of strength. Any one who has witnessed the remarkable effects of warm liquids thus injected in Cholera collapse must feel that recovery would be certain, if the improvement could by any possibility be made permanent.

Our hopes for the future, as to the treatment of Cholera, lie, as I believe, in the direction of supplying to the body, directly, its lost animal heat. I have witnessed the happiest results from an injection of warm salt water into the bowels, assisted by hand friction of the surface with turpentine and chloroform, and the application of bags of hot salt along the spine: in cases treated in this manner, we may expect to witness cessation of muscular cramp, restoration of perspiration to the skin, with increase of capillary circulation, and finally, to reward our efforts, a return of the excretions of urine and bile ; when these reappear, all vomiting and purging cease, and our patient is almost cured.

After recovery, the contrast between the Cholera and Fever patient is as great as it was during sickness. The fever patient has been overworked for 9 or 15 days without a suitable supply of food, and when convalescent, experiences a complete exhaustion of strength that lasts for many weeks. The Cholera patient, on the other hand, has been prevented from working, by constriction of the capillary vessels, caused by the absorption of the cholera poison,* and feels, on recovery,

* Whatever this may be, its period of incubation is 49 hours; that of strychnine is 22 minutes.

much like a man that has been half drowned, while the Fever patient resembles a man that has been half starved: the one is able to return to his work in the course of a few days, the other, only after the lapse of as many weeks.

There are two popular superstitions prevalent among medical men respecting Nature, which yearly slaughter hecatombs of victims; viz., that Nature is simple in her operations, and beneficent in her intentions; she is often both simple and beneficent, but at other times she is unquestionably both complex and malevolent.

An Egyptian fable informs us, that the votaries of Goddess Nature were divided in opinion as to whether she was transcendently beautiful, or hideously ugly; and that, in order keep up this difference of opinion which suits her purpose, she always wears a thick veil over her face.

> "For, with a veil that wimpled everywhere,
> Her head and face were hid, that mote to none appear;
> That some do say, was so by skill devised,
> To hide the terror of her uncouth hue
> From mortal eyes that should be sore agrised;
> For that her face did like a Lion show,
> That eye of wight could not endure to view;
> But others tell that it so beauteous was,
> And round about such beams of splendour threw,
> That it the sun a thousand times did pass,
> Nor could be seen, but like an image in a glass."

Before trusting Nature in the matter of Cholera, and proceeding to help her, it would be well to inquire whether she intends to cure the patient by her evacuations, or to put him into his coffin. For myself, I greatly mistrust her, and would wish to ask, previous to assisting her, whether she is really my Mother, or only my Stepmother. Our experience in Dublin has shown, that no more effectual mode of shortening life could be devised in Cholera than the "eliminant" treatment; and it was accordingly abandoned as soon as tried in that city.

It is much to be regretted, that an authority so deservedly held in high repute as that of Sir Thomas Watson, can be now quoted in favour of the treatment of Cholera, by the maxim, *similia similibus curantur.* So far as Dr. Watson has informed us, his change of opinion rests upon the statements of others, and not upon his own experience. He has suddenly become an advocate of the castor oil, rhubarb, calomel, and eliminant treatment of Cholera, and writes as follows:—

"When I last spoke on this subject in these lectures, I stated that the few recoveries which I had witnessed had all taken place under large and repeated doses of calomel, but I could not venture to affirm that the calomel cured them. At present, I am much disposed to believe that by its cleansing action, the calomel may have helped the recovery; and after all that I have since seen, heard, read, and thought upon the matter, I must confess that in the event of my having again to deal with the disorder, I should feel bound to adopt, in its generality, the evacuant theory and practice."

Sir Thomas Watson omits to add, that the cases here referred to were only six in number, of whom three died, and three recovered; which is exactly what might have been expected if he had not interfered at all.

Cholera from Bengal visits these islands, at intervals of about 17 years, and it is much to be feared, that on its next outbreak hundreds of patients will be sacrificed, in obedience to the dogma that asserts it to be our duty to assist Nature.

C. *Diabetes mellitus.*—This disease furnishes us with one of our best proofs that all the chemical changes, by means of which work is produced, take place in the blood and not in the tissues of the body; and, at the same time, an examination of its phenomena explains satisfactorily the regimen and diet which has been found, by experience, most suitable to the diabetic patient. I shall illustrate the disease by a case which was placed under my controul, by Dr. Stokes, some years ago.

A young man (æt. 20) named Murphy, suffered from fever (Enteric ?) in November, 1859, and on recovering, became diabetic; he was admitted into the Meath Hospital, in October 1860, where he remained, under my observation, until his death on the 12th January, 1861.

He was allowed, for nine weeks, to eat as much as he liked of certain kinds of food, which were varied, week by week, to suit his wants, my object being to obtain, if possible, the natural constants of the disease, undisturbed by external interference; the only medicine used by Dr. Stokes's order being opium, to produce sleep, and a little kreasote occasionally, to promote digestion. As the details of this experiment have been fully published, I shall confine myself to the final results. His food and excretions were analyzed from week to week, so as to determine the total quantities of sugar-forming and urea-producing food, as well as the sugar and urea actually excreted.

During six of the nine weeks, the sugar excreted was in excess of the sugar ingested; and the means of the whole nine weeks' daily excretion and ingestion of sugar were,

Sugar excreted, ...	9773 grs.
Sugar ingested, ...	9321 „
Diff.	452 grs.

During two of the nine weeks of observation, the urea excreted was in excess of the urea ingested; and the means of the whole nine weeks' daily excretion and ingestion of urea were,

Urea excreted, ...	1182 grs.
Urea ingested, ...	1349 „

The foregoing facts illustrate strikingly one of the prominent symptoms of Diabetes, viz., the canine appetite; the quantity, both of sugar-producing and urea-forming food consumed is more than double what is necessary to maintain a vigorous labourer in perfect health. An examination of the excretions explains the other prominent symptom of Diabetes; viz., the complete prostration of strength in the patient, notwithstanding the great amount of food consumed.

In a state of health, food produces three excretions only, viz., urea, carbonic acid, and water; in Diabetes, the farinaceous foods appear in the excretions as sugar, and not as carbonic acid and water; and the work necessary to maintain animal heat must be provided altogether at the expense of flesh food, which is the very form of food least fitted to maintain it.

The Diabetic patient resembles a racing steamboat on the Mississippi, whose supply of coals is exhausted, and whose cargo furnishes nothing better than lean pork hams, to throw into the furnace, to maintain the race. It cannot be wondered at that our poor patient, under such disadvantageous conditions, fails to keep in the front.

Let us compare together the minimum of work necessary to keep Owen Murphy alive, with the work actually supplied to him by the food digested.

1. I have already stated that Dr. Ranke found 660 grms. of carbonic acid excreted daily, in the extreme fasting condition, when he weighed 67 kilos. Now, since

$$660 \text{ grms.} = 10185.35 \text{ grs.}$$
$$67 \text{ kilos.} = 147.71 \text{ lbs.}$$

we find 69 grs. per lb. of body weight, as the minimum excretion of carbonic acid, consistent with continued life.

This quantity of carbonic acid represents a work generated by its production that would lift its corresponding pound of body weight through a height of

$$69 \times \frac{6}{22} \times 8080 \times \frac{9}{5} \times \frac{772}{7000 \times 5280} = 5.716 \text{ miles.}$$

Under ordinary conditions, the greater part of this carbonic acid and work is produced by the digestion of farinaceous food; but since, as we have seen, the farinaceous food is excreted as sugar in the Diabetic patient, and therefore does no work at all, the whole of the foregoing work must be done by the digestion of other kinds of food.

I have already shown that it follows from Lavoisier's experiments (confirmed in a remarkable manner by those of Regnault), that the work done by the combustion of carbon in the body is to the work done by the combustion of hydrogen in the proportion of 9068 to 3024, almost exactly 3 to 1; hence we have the work done, by Owen Murphy, as a minimum in health—

Due to carbon, ...	5.716 miles.
Due to hydrogen, ...	1.905 „
	7.621 miles.

This result is somewhat in excess of the truth, for the same reason that the calculated *digestion coefficient* of proteine is in excess of that found by Frankland from experiment; for the combustion coefficients of carbon and hydrogen, in organic compounds, are slightly less than when free. If we are permitted to reduce 7.621 miles in the same proportion as in the digestion of proteine, viz., 48 to 43, we shall find—

Owen Murphy—minimum of work consists of
body weight lifted through, ... 6.83 miles.

Let us now compare this minimum with the work actually performed by him when suffering from Diabetes, by the digestion of flesh food and production of urea.

2. I have already shown that the work produced by the formation of 501.28 grs. of urea is 704 ft. tons, by calculation from the composition of proteine and urea; this result should be reduced in the proportion of 48375 to 43155, in order to obtain the work given by Professor Frankland's experiments.

Making this reduction, we find that 500 grs. of urea correspond to 626.3 ft. tons of work, or 100 grs. urea to 125.26 ft. tons; or, in other words—

Every four grains of urea excreted correspond to five tons lifted through one foot.

Owen Murphy excreted, on an average, 1182 grs. of urea, daily, during nine weeks—which, by the foregoing rule, are equal to

$$1475 \text{ ft. tons} = \text{Murphy} \times x;$$

where x represents in miles the height through which the patient could be lifted by the work done per day; and is equal to

$$x = \frac{1475 \times 2240}{93.56 \times 5280} = 6.69 \text{ miles}.$$

This result is almost exactly equal to that already found as the minimum consistent with continued life, and explains in the most satisfactory manner the complete prostration of the patient, notwithstanding the consumption and digestion of more than double the usual quantity of flesh food.

In corroboration of the foregoing conclusion, I may mention that Murphy's temperature was found to be constantly 2° F. below that of other patients (chronic) placed in the same ward, and, in other respects, under similar conditions.

His unfavourable symptoms (so long as his powers of digestion were not impaired) were invariably alleviated by the free use of flesh food and fat, the latter being, instinctively, preferred by him; so much so, that during the delirium that preceded his death for 24 hours, he raved incessantly about "fat, roasted fat, which the angels of heaven were preparing for him."

I have studied many other cases of *Diabetes mellitus*, and found similar results in all; but I feel it to be unnecessary to describe them, as one well ascertained train of phenomena, carefully observed and recorded, is quite sufficient to establish the order of nature.

Conclusion.

I have, now, Mr. President and gentlemen, to apologise for the length of time during which I have spoken, and to thank you for the patience with which you have listened to me. I am well aware how much I am indebted to your kindness, for I laboured under two serious disadvantages in addressing you: in the first place, I had undertaken a task beyond my strength ; and again, my address is made, shortly after you had, like myself, been charmed and instructed by the luminous, learned, and eloquent oration of Professor Rolleston. I felt confident, however, that I possessed one advantage that he did not; I was a stranger in Oxford, and believed that my faults in matter and style would be leniently criticised; in this expectation, I am happy to say I am not disappointed; and again I thank you for your kindness. Two other advantages I share with him, which have contributed to his address as much as to my own—a profound respect and reverence for all honest labourers in search of truth, whether they have preceded us by 20 years or by 2000 years; and an unwavering confidence and faith in the future that lies before the Science of Medicine. We traverse a sea, mapped with imperfect charts, but assured of a safe guide in our compass and stars; but we cannot afford to neglect a single rock or shoal, buoyed for us by the skill and care of those that have preceded us. Let us follow their example, and mark with conscientious care, for our successors, the dangers we ourselves discover and escape.

Assembled, as we are, within the halls of the University of Oxford, the centre and heart of all that is intellectual and religious in the life of England; an University that borrows its accurate Logic, as well as its refined Ethics, from the lips of Aristotle; that reverences Euclid as the fountain and source of its elegant Geometry; and sits at the feet of Homer, Pindar, and Eschylus, to learn its poetry; we need not fear that Hippocrates and Galen will ever want admirers and students; but the Oxford of to-day has taught us, what many did not anticipate, that she is equally ready and skilful, as she has proved herself to be in cultivating Literature, to devote her vast intellectual energies to the encouragement and developement of the Natural Sciences, based upon the solid, and only permanent foundation of Mathematical research. The efforts made within the last few years by Oxford, to encourage within her walls the Mathematical and Natural Sciences, have won for her the respect, and warmed towards her the hearts of all that search for truth in the study of Nature. Our brothers in Oxford, like the Athenians at Syracuse, have gone on board the fleet, while we watch them from the shore, sympathizing in the sea fight; as they win, we shout; when they fail, we weep.

Long may the union of the far distant, but never to be forgotten Past, with the living Present, that now exists in Oxford, continue. No science, no profession, can benefit so much by it as that of Medicine.

THE END.

www.ingramcontent.com/pod-product-compliance
Lightning Source LLC
Chambersburg PA
CBHW021429090426
42739CB00009B/1414

* 9 7 8 3 3 3 7 2 0 1 4 4 9 *